The Pyramid of Love & Gratitude

& The Laws of the Universe

The Pyramid of Love & Gratitude

& The Laws of the Universe

Melynda Pearce

The Pyramid of Love and Gratitude

& The Laws of the Universe.

By Melynda Pearce

Edited by Chuck Gillespie

Formatted by Carole D. Fontaine

CONTENTS

INTRODUCTION

The *"Pyramid of Love and Gratitude"* was written to bring awareness of the power of the mind, and to understand the true meaning of love and gratitude. It is the fascinating journey of the creation of a stained-glass pyramid, created within the supernatural phenomena of a realm that many doubt its reality. This book is intended to elevate readers to a more advanced level of thinking and behavior to improve all people and their realities by utilizing the unlimited power and energy of their minds. It will illustrate how we can improve our realities by a series of actions, as a network of energy working together in harmony. This book specifically is written to honor the African-Egyptian people and everyone desiring to awaken to the true reasons for the human existence. Prepare to open your Heart and Mind to all the possibilities of your new future!

1

THE MISSING LINK

For decades, my personal reality was one of confusion. I awoke each day feeling that something profoundly important was missing in my life. When I would catch my reflection in the nearest mirror, I found myself constantly asking myself the same questions, "Why was I ever born?" and "Why am I even here on this planet?" Finally, one blessed day, I began to understand my life. A strong realization occurred to me, and I knew that I was meant to break through the painful memories of childhood trauma. There had to be relief from the damaging effects of anger and violence that colored so much of my early life, even to adulthood.

I managed to live day-to-day without being taught the meaning of a certain four-letter word…the all-important word. I felt that my life and reality were deprived, within a small and dark universe. I knew that I was void of the most important lesson I needed to fully live. Of course the word is simply...LOVE. It represents the primary, fundamental emotion. It was missing from most of my life.

Somehow, I had built defenses and hardened my world and survived my troubled youth experiences. I grew to adulthood. I became a mother of five children. I was determined to treat them differently than I was treated as a child. Unfortunately I had no role models or guidance to ensure any success. My life never prepared me adequately. I attempted to show them love and affection as best I could. I cared for them with all my heart.

Meanwhile, my life, beyond the children, continued to be a muddle, full of turmoil, leaving me to wonder why I always found myself in unhappy situations, surrounded by people who seemed to exist in a common, negative reality. Because of that type of existence, I never stayed in a relationship for very long. My several failed marriages that led to divorces were the unfortunate result. As I neared middle-age, I began to experience new, strange feelings. I was lost and confused. I didn't understand the meaning, but I sensed that something vitally important was right there in front of me, if I could only see it! What was it?

This melancholy was a constant presence in my life. This unseen, non-violent presence seemed desperate to remind me of something. What? My only brief escapes were through sleep and consuming alcohol. After drinking several glasses of wine, the intrusion of that un-named and unwelcome energy would temporarily evaporate. For an all-too-short while, I would be temporarily free from what seemed like it's wanting me to wake up and remember my obligations! My son Charles watched my nightly despair, telling me to face whatever it was. One morning he told me that he worried that I was killing myself with alcohol. He told me he loved me and walked to catch the bus for school. I knew, that sunny morning, that he was dead right. At that instant, I spoke aloud. I said, "Okay, I surrender. I will do my best to remember what I am supposed to. I will need your help and guidance!"

It was a beginning, the start of my formative journey of self-discovery. I would take many paths, searching for whatever had been absent in my life. All new directions seemed to be dead ends. I made little, if any, progress, or so I thought. In time, however, I began to develop an awareness. Something deep inside of me, a kind of knowledge, once fully known, had been lost and forgotten. I could literally feel it inside me, yearning to be discovered anew. I sensed that it was more important than my own life. If you can, imagine what it would be like to search for something you cannot describe or name, yet you know it is an intrinsic element of your very purpose and being.

I had to do…something. I followed my misguided logic and left my career in cosmetology, something I absolutely love doing! I turned to real estate, thinking that a fresh new perspective on work and changing my workaday environment might lead me to find some answers. It proved worse than futile. I began to feel more despondent and at sea than I had before. I then felt compelled to move away, to somewhere less urban and more pastoral, perhaps. So, I packed my possessions, sold my house, and moved. For a while there was some relief. Soon, however, the newness wore away and the shadows of dread reappeared.

The chances that I was falling into some sort of clinical psychosis were very real now, and my fears increased.

One positive result of my relocation into a more rural community was that now I was living closer to my beloved uncle Bubby and his family. He was my father's older brother. He was married with grown children, none of whom I had seen for over 25 years. I wanted to reconnect with them and I planned a visit. As fate would have it, the re-acquaintance began at the funeral of my cousin Connie, who had just passed days before. It was tragically strange that I was going to see everyone again, after so many years, under such sad circumstances.

The last time I had seen any of my cousins I was only 16. Now I was a 42 year old grandmother! So much life had passed in that time. I didn't want to attend the funeral alone, so I brought along my daughter Tiffany and her young son Robert.

Shortly after we arrived, I spotted a beautiful girl looking at me from a distance, as I was looking at her. She bore a close resemblance to my cousin Donnie, the sister of my deceased cousin. When we came face-to-face, I realized that it was not Donnie, but her daughter Tammy. The resemblance was uncanny. I was glad to see Tammy, and I hoped to see her mother as well.

When the service ended, Tammy and I spent the next several hours talking and sharing our life experiences. This young woman was only 32 years old and so warm and kind. I felt free to unload my pent-up struggles. I spoke frankly to her, confessing to having constant, strange feelings of an unseen entity of energy around me, trying to communicate with me. I admitted that I was fearful of losing my mind. Tammy immediately reassured me. "You're not crazy," she said, "You need to get the audio version of "The Secret" by Rhonda Byrne and listen to it." I had never heard of it. When I asked what it was, all she told me was "Get it, and listen to it. It will answer all of your questions." I purchased the CD version of "The Secret" when I got home.

Three days later, on a peaceful afternoon, I sat down and began to listen. Soon I started to have realizations. Like a flashing series of snapshots, my past came racing through my mind. Slowly I began to understand. I could clearly see how my state of mind could bring unfortunate situations to me, resulting in a negative reality. I understood that everything good in life, and everything bad in life, were of my own creation. Finding the proper reaction to negative situations, not of my creation, opened fresh channels to my higher self. I saw my own, real ability to use freewill to make choices and affect change in my life. I found a process to work through my experiences.

Although some periods of my strange feelings remained, the intensity, overall, was gradually lessening. I continued to listen to "The Secret" CD as often as I could. Within several months, I was feeling stronger and stronger. My reality absolutely was improving! I knew that my mind was expanding into the light of a pure reality. I kept listening. I continued to reach out for that missing something in my life.

One day, silently, quickly, and effortlessly, I understood the cause of all my anguish and doubt. The once bad cells of doom and despair that colored my reality for many years had converted into cells of love and light. It was a major breakthrough! I began to understand purpose in my life! I began to cry uncontrollably, as a strong surge of energy shot through every inch of my body. I felt a moment of sheer bliss. I was, for the first time in my life, complete. All those years of unknowing and searching for a missing truth, so simple, yet so profound. At that instant, I realized it was my lack of believing in, and loving, ME. How obvious! How necessary! My life would be forever changed from that moment forward.

Finally, I understood the connections to the most important part of who I am, and who I was in past lives. The darkness of the world had lifted off my shoulders. I was now free from any influence of a reality created by mankind, whose instinctual actions are based on limitation, chaos, and hatred. I was able to discover the forgotten knowledge of my past, and my purpose for the future, all safely protected in my cells. From that moment I fully understood my missing element of Love. My reality strengthened and grew at a continual pace. My energy vibrated strongly. My newly-healthy reality improved as a result of my constant vision of love and peace for all people, from all walks of life.

NOTES

2

DREAMS

Not long after my life-changing realization, I had an extremely vivid dream. It was so realistic! It replayed over and over through my head the next morning as if I had been front row center in a theater, watching it being performed on a stage. I couldn't wait to tell my daughter Tiffany later that morning.

In the dream, as I began to explain it to her, a clear voice from some kind of shapeless energy spoke to me. I was instructed to create a special, spiritual object. It would be shaped as a pyramid. Its purpose will be to "honor all Africans from the past, present, and to eternity." The voice said the colors are to be black, white and gold. The black must start at the bottom and go up at an angle to the top. The white with the gold will follow. When finished it will be forever known as the "Pyramid of Love and Gratitude."

Those were the complete instructions. There was no other communication or guidance as to material or size or whatever. Without a moment's hesitation I felt I had to make it happen, but how?

Three days past and then, that night, I had another dream. This time when I woke up, I had no recollection of specifics, but I was left with a sense of its importance. The next night, the same dream occurred. This time I tried, in my sleep, to make a mental note to somehow remember it in the morning. However, the next day it was lost again. I knew I had the same dream, but I just couldn't remember it. Right away I promised the messenger that spoke to me, to please give me one more chance and that I would wake up and write it down.

That night the dream returned. This time, as soon as the voice was finished speaking, I woke up. I glanced at the clock. It read 3:37 a.m. I hurried to the kitchen where I had placed a notepad before going to bed. This time I was ready to record the dream. The words flowed out of the pen to the paper…"Only when you open your heart to the love that has been waiting for you in the universe, only then will you understand the true meaning of life and living." I laid down the pen with a slight tremble in my hand. In the quiet stillness of a dark night I knew that I never was, and never will be alone.

Within days, with my daughter acting as a sounding board, I started to brainstorm ways to physically create the pyramid. We asked several building-craft folks for ideas too. Nobody would recommend a suitable construction material that resonated with the design. While I was stalled there, I felt compelled to register the concept with the U.S. Trademark Office.

After the registration process, I sort of had a mental block about the project. I really was unable to conceive a practical way to actualize my dream. So, for the time being, I sort of tabled the "nice dream" idea and turned my attentions back to my work. Within weeks a major political event began. A certain junior senator from Illinois announced his candidacy for president. Barack Obama had entered the presidential race. Tiffany and I wondered if my "dream" had any connection to the coming historical event in American politics. It was too premature to believe there was any connection. Later, that question would be answered as history played out.

It was no coincidence I reasoned. I was amazed by the timing. For the first time in my life I was truly believing in myself. I felt exhilarated. I was in control of my life, probably for the very first time. There had to be some connection. I started to search for answers. I had never really enjoyed reading, but now, I was reading voraciously, looking for answers with a specific purpose to share with the world.

Not long thereafter, I became inspired to return to live in my home town of Orlando, Florida. My husband (at the time) wanted to stay away from the city, in the small trucking community where we were living. I acquiesced for a while until one day, a man appeared at the front door. In his hands were legal documents from Polk County, Florida. The papers were to inform me that the owner of the house we were now renting was in foreclosure. The mortgage payments were past due.

This was bad news for the investor, but good news for me. I packed my things and left prior to being forced to leave the property. My husband insisted upon staying in the area though, and rented a condo. For me the decision to leave was easy to make. I decided to move back to Orlando and rebuild my life, eventually parting when I discovered my husband's philandering ways.

Back in a suburb of Orlando, where I grew up, I began to rebuild my business as a cosmetologist. I moved with two children in tow into my parent's home and began to reassemble my life. Each day I became a bit stronger and more aware of not just my inner vibration, but also the vibrations of those around me. I promised to keep in touch with my second cousin Tammy, whom I credited with saving my life by steering me in the direction that I was so desperate to find.

One weekend Tammy and I were off on a journey inward together. We met at the Cassadaga Spiritualist

Camp in nearby Lake Helen, Florida. Since the late 1800s, this unusual community has been a sanctuary where seekers, faith healers, and others, searching for insight and guidance, gather. We planned to attend a spiritual church service and hopefully, receive a free psychic reading that afternoon. It happened just as we imagined. I was chosen by a medium named Diane, who offered to give me a free reading. I readily accepted her offer, while keeping a skeptical mind.

Diane began the session by telling me that I was "a kind spirit and that I enjoyed working with colors." Tammy and I looked at each other in surprise because I worked with color for a living, as a hairstylist. Soon the psychic just stopped her deliberate, soft explanations and blurted out loudly, "stained glass!" I was taken aback a bit and looked around at those around me and sort of shook my head a bit, showing that I didn't understand. Diane quickly apologized for the outburst and admitted that she didn't know why she said it either.

She continued my reading and was soon relating impressions of my childhood. She told me that when I was a little girl, I went in a different direction than everyone else. As I drew closer to hear her speak, she did it again. She suddenly stopped and shouted out, "Stained Glass!" What the heck lady. I thought to myself. Is this part of some kind of show? Again she apologized and insisted that she still had no idea why she repeated it.

Once more she began to talk about my life. It was fascinating. In a short while it was over, without any more shouting, and she got up to move to the next person. Before she moved away though, she turned back to me and said, "Honey, I don't know what you've been doing lately, but keep doing it. You're on the right path!"

Her words were galvanized in my mind. "You're on the right path" played over and over in my head. The "stained glass" interjections were interesting, but I didn't attach any significance to them, at least at that time. All I could think about was how wonderful it was to believe in myself and exist in a new and improved reality by loving more and thinking properly. That psychic's insight into my world was like a shot of affirming serum. It was a booster for my soul.

NOTES

3

THE BUILDING OF
THE PYRAMID

Weeks past after my spiritual encounter with the psychic. I was keeping busy getting settled in at my mom's house. I found work too. It was, in the vernacular of the hair world, at a "chop shop". In other words, it was a low-budget, high-volume, hair salon. Unfortunately, the energy among the employees was very dour and withdrawn. There was very low morale probably because of the low wages, long hours, and inept management. I soon grew very weary of all the infighting among the workers. My whole life had been spent around family and friends that were in that exact type of fatalistic existence and mindset. I wanted no part of it ever again.

After four months of this mind-torture, I packed my combs and shears and left to work in a two-person studio. My new boss was like a breath of fresh air. She was stable in her beliefs of the Cathology faith. I could sense too that she was in a favorable vibration of total success. It was the type of quiet, work environment I needed to be around. I recognized her personal financial burden and I was happy that my booth rental fee was helping her meet her obligations. It was easy to settle into the daily routine and let the memories of the spiritual camps and psychics slowly fade over time, until one fateful afternoon.

It was about three months after beginning to work in the small shop. As I was leaving for the day, I noticed a window sign on a door a short walk away. It was a flyer advertising new classes in creating stained glass, starting soon. "That sounds like fun," I thought. I made a mental note of the posting and hurried home to dinner. Several weeks later, when I had some spare time, I decided to stop in and check it out.

I walked into the stained glass shop only intending to stay a few minutes and ask about the classes. I certainly didn't anticipate what happened next. After barely taking my third step through the doorway, it began to hit my conscious mind, in waves. I immediately was focused on the dream, the voice, my instruction, The Pyramid! I don't know how, or why, but I saw the stained glass displays and right away I knew it should be the materials to construct The Pyramid. That sudden realization was a surge of energy through my body, not unlike the sensation I felt that moment when my reality had improved and changed for the better all those months ago! I took a deep breath to steady myself and then asked the man behind the counter the question I had needed to ask since I met the shouting psychic.

"Sir, hell-o, uh, I was just wondering, can you make me a pyramid out of stained glass? My words coming from my mouth came as if they had been pent up inside me for a long time. I felt like I was finally, at long last, breathing life into something that had been a part of me for months, or maybe eons. I felt relieved, excited, and happy.

He smiled at my request, nodded, and spoke, "Do you have a sketch?"

"Well, no", I responded.

I had only registered the pyramid with the patent and trademark office, then daily life clouded my task, until just now. I did remember the voice, and the details about color and design. How could I ever forget that! I clearly described the design as best I could. I asked how much would it cost. Again he smiled, scratched a few notes on a piece of paper, and said, "350 dollars." My first reaction was a simple, silent "ouch", kind of expensive I thought. But I had to have it made, and it had to be stained glass, and the surge of energy, the flood of Pyramid thoughts… This was right and this was the place! After just several seconds I smiled and heartily agreed. The shopkeeper was emphatic that he really needed a sketch to work from, and to bring it to him when it was drawn. He'd have a look at it and see what he could do. I agreed.

Just a few moments later, as I turned to leave the shop, a very unusual thing happened. It was as if something or someone assumed my mind and body and began to channel and speak through me to the man, through my voice, my mouth. I heard my voice saying as I turned back to the man, "Sir, how would you like it if someone took you from your home and made you to do things against your will? Then, eventually, you would gain your freedom, but find that people did not accept you as you are?" I went on, the words pouring out of my spirit. "Vision is a gift from the Universe. If everyone was struck blind there would not be any racism in the world, because you cannot hate what you cannot see."

The shock of my words reverberated for several seconds in the air. I felt as if I was in a kind of waking paralysis. Something or someone unseen, without shape or form, spoke through me. The befuddled man stood quietly, slowly nodding, staring at me, wondering, no doubt, if I was more than a little off-center. I didn't care. I knew my utterances were true, simple and unadorned. I simply smiled, turned, and casually strolled outside. The evening Florida air was welcome and refreshing. I was empowered. I was off to buy art supplies to attempt to sketch the object of my obsession.

After a brief stop at the corner store of happy and healthy, gathering the necessary supplies, I headed back to the now-closed salon. I knew it would be quiet there. I could work undisturbed. When I entered the darkened room, I flipped on the light switch and locked the door behind me. I went to the break area in the back and poured my purchases out onto the table. I sat down and opened the new sketch pad and picked up a soft-lead pencil. I immediately started drawing. It was as if I had been an artist my entire life. I knew the exact dimensions that the pyramid required. It was as close to "automatic writing" as I had experienced by then. It was exhilarating.

Within minutes, I had completed the picture, even coloring each section with the appropriate colors. I didn't have to start over or erase a thing! Although I had no previous design plan, the image flowed from my pencil to the page. My mind recorded and stored in my cells the proper angles and how the three colors would intersect.

My eyes scanned every inch of the completed color sketch. I followed the lines and the angles. I visualized how the black, white, and gold colors would appear when interacting with light and shadow. I realized I had captured the look. While I was creating the sketch, my daughter Cristia dropped by the salon for a visit. She saw the pencils and the sketch and wondered what I was doing. I explained the Pyramid was finally going to be created. She was so excited! While she stared at the sketch for a minute she suddenly said, "Mom, I think you should add a star at the top."

Yes I thought! Could this be the missing other part of my missing information? It felt so right. Immediately I sketched a star for the top of the designed sketch. It was complete.

I knew this couldn't wait. I gathered up my things, we hurried out and locked the salon. I hugged my daughter and thanked her, I rushed back to the man that was going to make my dream a reality. He was standing behind the counter looking at his computer screen when I breezed through the door. He gazed up at me, "Oh hell-o again. "You know kiddo," he said, "I think I was going hard on you with that price. I've just been thinking that I'm only going to charge you $175."

Wow! I was thrilled. I hadn't really thought about if I could afford this or not. I was just acting by impulse. Like everyone, I had bills to pay and I had a limited income. Now, at least, this wouldn't break the bank. Why had he changed the price, really? Did my little "outburst" earlier move him in some way? Regardless, I was grateful. I agreed to the price immediately. I brought the sketch out of the sketchbook and he immediately began the process of scanning it into his computer. He said he would give me a call in a few hours. I left the shop feeling that terrific. Another significant piece in my life mosaic was being put into place.

Back home, a couple hours later the phone rang. The artist said that he had loaded the image to the computer and a digital rendering of the finished product was ready to be shown. I told him I'd be there in a flash. I grabbed my purse, punched out the light and hurried out into the night. In just minutes, there I was, standing in front of a 19-inch computer monitor, staring transfixed at the black, white, and gold, three dimensional image. I could see exactly what the pyramid was going to look like in this electronic form. I was awe-struck. I started to cry. I don't know why I was crying, but I was filled with over-flowing joy. I knew, at that moment, that this creation would be more significant than I could truly describe.

The artist assured me that the construction work would begin right away and the finished piece would be available in only a few days. I left the store as if I was floating on cloud nine. I was exhilarated, feeling my spirit blissfully enjoying every moment of this love and gratitude experience.

When I got back home, I called Tammy and told her that the pyramid was finally realized and being created. I shared with her the entire sequence of events. She reminded me that this was what the medium had been talking about.

"How could that have slipped my mind? Of course it was!" It all was astonishing. I was more convinced than ever that I was headed in the right direction. I had no idea what was coming next or what was about to happen after the pyramid was finally coming together and being created into something tangible that I could treasure. This had been such a remarkable day! Little did I realize that what would happen, so unexpectedly, later that night, would be the most incredible experience of all!

NOTES

4

ANOTHER DREAM

That night, as I was falling asleep, a sense of well-being and bliss came over me. I knew I was exactly on track, as the medium had foretold. In just days, I would have the physical evidence of my beyond-normal experiences. As my sleep grew deeper, I experienced an extraordinary dream.

Flying toward me at amazing speeds, a beautiful angel appeared. As she passed, our eyes met. Abruptly she stopped in mid-air. Her long, beautiful curly hair flowed with each movement of her energy. She hovered slowly above me, her torso curving and arching for a moment, then she was moving toward me, in a floating, seated position. Suddenly, two beings appeared. The angel quickly looked away from me and began speaking to the new apparitions. They were seated on a small cloud. They appeared to be identical, like twins. I looked to my sides and noticed a boy angel on each. They escorted me as I began to witness the angel and twins.

Instantly I had a semi-conscious thought. Was I dead? Is this the next experience beyond? I was frightened. Had I escaped life without the chance of saying goodbye to my children and grandkids? I wanted to wake up. Now! I jerked my head, violently from side to side, trying to force consciousness. Thank God it worked! I awoke in a cold sweat. My heart was racing but that was okay, I was very much alive. What had I just gone through? What did it mean? The three apparitions seemed to be trying to communicate with me, but I was too terrified. What did they want me to know? Will they come back and try again?

The next day at work, I had a new client in my chair. As I worked on her hair, I found myself sharing the details of my dream. She listened as if she understood some meaning of the event. When I finished she thought a minute then offered her explanation of my dream. She said that she believed the angels were trying to deliver me a message, but my fear and recoil stopped them before they had the chance. Maybe she was right, I thought. The more I thought about it, the more I was sorry that I woke up.

My client could see how distressing the whole thing had been on me. In a reassuring voice she told me not to worry, adding that perhaps the "higher realm of all realities" was pleased with my making the pyramid. Maybe the angels were simply there to say thanks. "I think it's likely that they will visit you again" she offered. "It's probably a good idea to pay close attention to everything you might experience."

Everything she told me seemed to make perfect sense. I was so grateful for this lady, a total stranger, coming in, hearing my tale, and giving me clarity. She was so friendly and engaging, I hoped I would be able to share this story with her in the future. I got her phone number and she left the shop.

She never returned. In fact, when I called her phone number a few days later, the recording informed me that the number did not exist. Could she had been a physical angel that needed to bring me the message that I pleased the Angelic realm of existence? Only in time would I understand the reasons for each encounter I would have and for what specific purpose.

Whatever the cause of the dream and the message, the next several nights were dreamless, but very restful. I thought often about the pyramid construction, hoping I would hear something soon. In a few days the call arrived. The Pyramid was ready. I could come by and pick it up today or tomorrow. Of course I couldn't wait a second longer! It was as if I was picking up my child, my creation. This Pyramid was a living part of me now. It was conveyed to me by a powerful being from another dimension that contacted me in this realm of consciousness.

I grabbed my keys and rushed to the truck. I drove as if I was being guided by a kind of "Star Wars" tractor beam. It was as if the truck was driving itself! I parked the truck and rushed to the shop. As I walked in the door I immediately saw the finished Pyramid sitting alone, in the middle of a table, the stained glass glistening through the black, gold, and white. I slowly walked closer, taking in its splendor.

As my eyes surveyed every inch of the creation, they began to fill with tears. Like waters of life, flowing through a brook, they streamed down my cheeks. I was completely stunned by its beauty. I began to imagine all it would be. In the presence of it, I felt a raw energy. I realized that this was something very significant to the consciousness of the world. It was a message from the Divine himself, full of infinite intelligence and love, far more powerful than anything we could imagine.

I carefully bundled my exquisite possession and drove home. I placed *The Pyramid of Love and Gratitude* in a safe place in my bedroom. I really had no idea what I was to do with it now or why it even existed, besides to honor all Africans. I began to think. The timing of the first African president of The United States absolutely wasn't any coincidence either. The Higher realm had a strategic plan too. People and events unfolded simultaneously to gently take back the reins to awaken us calmly without stress or shock.

In the days that follow I decided to spread the word globally about the Pyramid. I launched a website and made a graphic representation of the Pyramid, placing it on the site as a logo. Over time the site was replaced but I am very proud of all the effort and thought that went into its creation. I also had to re-register the Pyramid, now that it was in a three-dimensional art form.

I soon felt that my next efforts on behalf of the Pyramid would be for me to seek truths about myself. Because of my being drawn to the Pyramid's creation, I wanted to know more about my past lives and their connections to Africa and its people. That's where I began my research. I also read all I could about slavery. Page after page of reading saddened my spirit at the thought of what had been done to the Egyptian-African people. It was difficult, but I self-taught myself more than I was ever taught in public schools.

I was intrigued by the slave trader's passage routes from Europe to Africa to America, and back to Europe. It formed a triangle. I felt strongly that there was a connection to my Pyramid, not just coincidental. The African slave trade formed a literal triangle in the ocean. I kept wondering what it was about the Africans. What caused them to be singled out, shackled in chains, sold as livestock for hundreds of years! I once believed it was the color of their skin. The more I read, the more I saw other more perverse and ancient reasons emerge.

I pored over the history texts about President Abraham Lincoln. I read how his first and second attempt to free the slaves in 1863 were unsuccessful. Why weren't they? Could he have made a self- discovery of the bonds connecting Egyptians, Africans and the higher realm? It was not until his third attempt in 1865, when General Granger personally delivered the emancipation proclamation to the waiting press. The Proclamation of Emancipation stated that the slaves were to be freed forever! The true meaning of that word "freedom" is still being sought over 150 years later, as is the true purpose of the descendants of America's dark, divided past.

NOTES

42

43

5

THE LAW OF ATTRACTION

The amazing events of the recent past brought many things into focus. I started to feel more confident in thinking and drawing from my own intuition. When I'd find myself thinking about the future, I often wished for a life-partner. I wanted someone to share this journey, I wanted to live the life I was meant to live. I asked for spiritual guidance and help.

I prayed for a partner, a friend, an intimate soulmate. I wanted someone brilliant, open minded, drug-free, and alcohol-free. I wanted a special person to love me completely as I loved myself. I promised that if that person didn't walk into my life, I would spend the rest of my life alone. I wasn't being flippant. I was simply dedicated to my commitment to never again be involved with the type of men from my past. Those relationships all failed.

I knew they were due to my bad choices and my once low self-esteem. I only blame myself, not the men from my past. I am fully responsible. But now, my eyes are opened wide. I will have what I want and need, or I will do without. I can be miserable by myself, and I can also be happy by myself.

Later in the year something terrific slowly began to unfold. One evening my eldest daughter Crista and a friend asked me to join them for an evening of drinks and dancing with the girls. What fun! Sure I'd love to go! Those days, I was doing little more than work and come home, day after day, without much recreation. It sounded like a great diversion.

The scene at the Why Not Lounge was loud music with strong drinks…Why Not? We were having a great time. Our conversation was light and full of laughter. At one point in the banter, I started sensing someone's presence very close by. I turned to look over my shoulder, but no one was there. I felt a different type of energy around me. It was kind of strange, but not unpleasant. Several minutes later I turned and looked again. This time, a man was there, looking directly at me.

He was an attractive man, just smiling and waving hello at me, sitting at a table a distance from us. I felt a little embarrassed and quickly looked away. In less than a minute, he left his table and made his way toward me. He smiled standing in front of me and asked me to dance. I didn't really feel like dancing with anyone but, my daughter Crista and her friends, urged me to accept. I was a bit leery of interacting with the men in a nightclub, I wanted no part of a "drinking man", usually the type of guy you meet in that type of establishment. I thanked him but declined his offer. After he walked away, I coaxed the girls to join me on the dance floor. While I danced with the girls, that friendly stranger danced near me and tried to strike up a conversation. He was fighting a losing battle though, because the music drowned out almost every word he shouted.

The girls and I danced a few more songs and then went back to our table. The smiling man, undaunted and uninvited, followed. He then invited us to join him and his friends at their table. We glanced at each other quickly and all nodded a short "okay". When we all assembled at the new table, our host introduced himself as Joseph. We ordered drinks and chatted a bit more, although most of the small talk was drowned out by the music. Joseph was pretty focused on me. That was obvious. I agreed to dance a couple fast songs with him before it was time to leave. When it was time, the girls and I left together and headed home. My spirit was light and airy.

The following weekend, Crista and I went back to the Lounge. When we walked in, Joseph was already there. He spotted us quickly and invited us to join him at his table, like last week. Once again the music was too loud to talk over, but unlike last week, I wanted to learn more about him. We walked to a quiet hallway, right outside of the lounge. Soon, we were chatting away. I asked him if he knew anything about the laws of the universe. He said he didn't. I proceeded to tell him about my Pyramid dream and all that had transpired within the previous year. He seemed very interested in all of it. After hearing my story he told me, "It's great to finally meet someone with something between her ears." He was giving me a compliment, I think. I took the comment as one.

We sat and talked the evening away. I continued to explain the laws of the universe to him, and how the mind is either negative or positive and sometimes in between, like I used to be. He told me about his job. The name of the employer sounded like some kind of marina. Later, I had a good laugh when I found out he was a professor at "Full Sail". It's a University in Winter Park, Florida.

Finally the night was ending and Crista and I got ready to leave. I gave Joseph my phone number and email. I felt good. Maybe this time, I'd met someone very special. I hoped that he might become the best friend and life-partner I had asked the Universe to send to me. I felt like I was beating the odds. I had met a man, in a bar, who didn't drink alcohol, smoke cigarettes, or ingest any kind of drugs. I felt that my divine request was granted. I truly believed I had finally met the perfect man.

The next Monday an email popped into my phone. It was from Joseph, just saying a quick hello. For the next four days, I was regularly checking my email and text messages, hoping for any new communication from him. By Thursday, I was still waiting for a message. What gives with this guy? Maybe I had misread him. I had to laugh at myself for being so quick to think that Joseph was actually directed to me in the first place, even though he was everything I had requested in the first place. I thought I was being kind of silly and put the phone away for a while. I certainly wasn't going back to that club ever again. I felt a little foolish, even giggly, that I was blithely cruising down the wrong path yet again. This was just another example of how my conscious mind could play tricks on me. I released all the anxious thoughts and went about my day. Several hours later, wouldn't you know it, my phone rang.

It was Joseph. He began very apologetically, telling me how sorry he was for not calling earlier. He said that his work hours had been just crazy this week. Regardless of how foolish I was feeling just a short time ago, now, hearing his voice, I was really glad he had called. He asked me out for a date night to the movies. Of course I accepted! When we hung up, I kept replaying in my mind how we met and feeling the energy I felt that night. It had been a good night.

There was something very strong and positive about Joseph's energy. I was drawn to him like a powerful magnet. It was clear from that day forward that our energies were in close harmony with one another. We started seeing each other often and soon, I was spending all my time with him, many days and evenings, falling in love. I was still living in my mother and stepfather's house, seeing them less and less because Joseph was becoming my life.

I could really see my life changing for the better. Until that time I had spent all my adult life drawn to the wrong men. I settled for second best I guess, always, because of where and how I was raised. With Joseph, for the very first time in my life, I was finally outside of that world and it felt wonderful! I was experiencing the positive side of life. Someone I had drawn to myself, due to my change in thinking, had helped create a brand new reality. This stuff is like magic! Every day brought more proof that this was the way life should be always. I spent all of my time with this wonderful man. I felt strongly that the universe sent him to me, and that he was chosen to share this magnificent path with me, guiding me through this magical experience.

Our relationship grew quickly. Not long after, Joseph asked me to someday consider moving in and living with him. At first I told him maybe in time, that we should wait because our relationship was so new and we had only known each other a short time. But after just a few months, I couldn't resist anymore. I decided to make the move and end living at my mother's home.

Although I loved my mom very much and always appreciated her help, I became unable to tolerate my sharing her house. I knew she was missing something inside of her that I couldn't provide. Living with her I heard almost daily complaining about everything under the sun. Those irritants she expressed I had no influence over, nor did I cause in any way. I was such a constant negative vibe! I knew she was only acting out, afraid of losing me and being left all alone. My mother was missing what I had missed my entire life. She needed Love in her heart. Her deeply wounded psyche, from her childhood forward, left a dark void in her soul that no one could heal.

Looking back to these recollections now as I write this book, one thing has become very clear. Once an individual has "crossed over from the negative side of reality", there is no tolerance for negative energy in their lives, none at all. All that darkness is to be avoided, from anyone, including your own family. It can slowly bring you down until you are drained completely. It's a slippery slope to self-destruction, creating a self- reality that becomes a burden to the spirit. Picture your mind as a kind of video camera. It is always ON and the red RECORD button is always pushed. It will manifest whatever it experiences.

I loved my Mother with every fiber of my being, but her mental "place" was somewhere I could never return again. She acted resentful and was constantly angry at anything. I knew that in order to preserve my happiness, I had to leave her to her personal torments.

I asked Joseph if his offer to live with him still stood? "Of course!" he said. I could tell he was thrilled at the thought of us being together. I was relieved and very happy. I was in love. I was determined to not allow anyone to destroy what I had accomplished or block the positive path I was traveling with Joseph.

NOTES

6

THE PYRAMID'S JOURNEY TO THE WHITE HOUSE

Moving in with my sweetie went great. We were together in the new home he had just purchased. It seemed natural and easy. From my new home and life with Joseph I began to regularly drop in on mom just to visit. It always seemed to make her happy. I always felt good inside seeing her smile and laugh. I loved those moments together. I always hold her memory close to my heart.

One day while we were just putzing around the house, Joseph sat me down for a serious chat. There was something definitely on his mind.

"I've been thinking a lot about the Pyramid. I think you need to have another one created and send it somewhere important, where it will be seen and appreciated. I think it's meant to be in the White House."

"Oh Joseph, what an interesting idea! I like it."

At first though, I was unsure if it would matter to anyone. Maybe it would just be tossed into a corner, or even rejected outright. The more I thought about it though, the more it made sense. We actually donate the second Pyramid of Love and Gratitude to the people of America, to reside in the "people's house", to honor The African-Egyptian people. Yes! This divinely inspired piece needed to be there!

For the next few days I was constantly asking friends and family members what they thought of the idea. They all agreed that it should go to Washington. My mind was finally made up. The second Pyramid was going to the home of America's first African president, Barack Obama and his family. Because I knew we were keeping the original, we needed a perfect duplicate created for The White House.

That Sunday afternoon we went for a drive. Several days before we had passed what appeared like a stained glass store and we decided to check it out. When we pulled up, we saw they were closed on Sundays, but we parked and spent a few minutes looking in the windows. We could see that it was a beautiful gallery. There were glass angels everywhere, with light streaming through their shimmering bodies. There were stained glass nature scenes and beautiful animals too. We just nodded to each other. This was the place!

Although we were definitely interested in getting the Pyramid project going, our other work-a-day lives sort of took over for a while. We were working, coming together at night, and thoroughly enjoying ourselves together. About three months had past. One evening, Joseph reminded me that we ought to get back to that shop and see about getting that copy created.

He was absolutely right. We were just letting the time get away from us. That Saturday, on our way to an afternoon at the beach, we pulled up to the shop. There was a sign on the door announcing that they would be open later in the day. We decided to not spoil a beautiful day and headed to the ocean. I would come back over by myself on Monday, my day off from work.

When that important morning arrived, I jumped up, showered, dressed, grabbed a cup of coffee, and shot out the door with my valuable package in hand. As I drove along, I went over in my head how I wanted the meeting with the new artist to go. I had no interest in revealing the true meaning and importance of the Pyramid. I believed that it could impact the price he might charge. I wanted to negotiate a fair, unbiased price this time.

When I arrived at the shop, the lady behind the counter looked up from her morning coffee and shuffled over. "Hi," I said. "My name's Melynda Pearce. I noticed from your windows that you have a lot of interesting stained glass pieces in your gallery. I wonder if you ever recreate this piece?" I unwrapped my prize, standing it on the counter, as it caught the morning light spilling in from the store's window. "I'm looking to find someone to copy this Pyramid, and maybe clean up some of these lead lines here, and here."

I showed her the blueprint too, but she said she wouldn't need it. She put on a pair of glasses and carefully looked over the Pyramid, from every side and angle. After a minute or so, she glanced up and said, "Yes, I'd love to copy this for you. For labor and materials, the fee would be $138.00. I can have it ready in two days. Okay?" I accepted the cost she quoted me without any bargaining. It was less expensive than the previous price. I thought that was a reasonable fee considering the first artist's fee.

"That sounds great," I said. "Let's shake hands on the deal."

We did. I was eager to try to explain why this piece was so important. I told her about my dream from two years ago. I explained that the Pyramid was shown to me as a symbol of unity, honoring African-Egyptians of the past, present, and to eternity. I was thrilled to inform her that the copy they would make was going to find a permanent home in Washington, in The White House!

Right then, at that instant, something miraculous happened. At the exact moment that I told the artist about where the Pyramid was heading, her husband, who had overheard our conversation, yelled from the other side of the room, "Honey, I told you I dreamed that our stained glass was going to be in the White House!"

"What is he talking about?" I said, in a kind of shocked and surprised voice.

The artist rolled her eyes upward and said that her husband had a dream several months ago. In his vision he witnessed a stained glass of their creation being placed into the White House. She said that her husband told so many of their friends and family about his dream that she finally told him to stop because he sounded crazy.

I knew it was no accident or happenstance that he had that particular dream. Her husband had received a very special gift of instructions, just as I had. He was able to feel the positive energy that Joseph and I imprinted months before when we stopped by the shop and touched their windows. I quietly smiled as I realized that we were not the only people connected to this mission and purpose.

I knew at that moment there would be more to come and something very good was behind this purpose and reason.

What an amazing day! I couldn't wait to tell Joseph about the events at the shop. When we were together, later that evening, I told the story from beginning to end, careful not to leave out the slightest detail. When I finished, Joseph was happy that the artist was engaged. I think he was a little skeptical about my interpretation of the artist's husband's dream and its significance. No matter, I was certain of its importance.

When the Pyramid was ready, Joseph and I went to pick up Pyramid number two. After introducing Joseph to the couple, I intentionally referred back to our conversation two days earlier. I asked the husband, "How does it feel that the dream you had of your stained glass going into The White House, is really going to come true?"

"It's a miracle. I can't believe it yet," he said. I watched Joseph's face for his reaction. I sensed that he believed that the man was speaking from his heart. There was no skepticism in Joseph now.

We stood around the counter chatting for a short while, but not too long. I was eager to get our artifact safely home. We didn't talk much in the car. I think we were both feeling the building of importance of what we were doing. After the twenty minute drive we got home.

When we got inside, we carefully unwrapped the packaging. When it was completely revealed, we couldn't help but stare at their work. The detailing was impressive. It was a beautiful piece of stained art. Joseph picked up the original Pyramid and sat them side-by-side. We studied them both for a minute, then Joseph spoke, "Honey, I'm sorry, but we can't send this Pyramid to the White House."

"What? Why?" I said.

"Take a look at this"

I leaned closer as Joe began to draw my eye to what he was seeing. Although the artist was very exacting in her attempt to duplicate the Pyramid, it was not identical. There were definite differences in the design. They were subtle, but specific. I felt heartsick, but I knew that it had to be an exact replica. The lines didn't curve, they were straight.

I called the shop. I wasn't eager to tell them that we were not satisfied, but I knew it had to be done. I carefully explained that her version was a lovely piece but not an exact replica of the first Pyramid that I had created.

I offered to return it and provide her with the blueprint that she initially refused. She did not want this one back because she had nowhere to store it, but did offer to give me a discount on the third pyramid. I told Joseph, and we agreed.

I drove back to her gallery. This time, I had the blueprint in hand. She looked over the specs and I could tell that she understood the changes necessary. She said that a little more time would be needed, but it would be completed in four days. She assured me that it would be well worth the wait.

As she instructed, on the fourth day we got the call and hurried back to the shop. Once again we saw a beautiful work of art on the counter for our perusal. Again, we looked over every inch, while comparing to the blue prints. The duplication, this time, was a perfect success, maybe even better than the original! The workmanship was outstanding. It was indeed worthy to be gifted to the President of the United States.

With Pyramid in hand, back home, we started to look at legal procedures for sending unsolicited gifts to The White House. Initially, the law states that advance notice must be provided. I hurried off a "letter of intent to ship", then waited for a response. After about two weeks, I decided that enough time had passed. I presumed that the letter had reached 1600 Pennsylvania Avenue by now. I wanted to be able to ship the precious cargo to arrive safely in time for Thanksgiving, 2008. Because I wasn't legally required to wait for an official reply of consent, I was free to plan the Pyramid's important journey.

The November, 2008 election had brought the most amazing outcome in American history. Barack Obama had become the first African President. Twelve months later, the time for turkey and cranberries was drawing close. On the Friday before Thanksgiving, I drove to my nearby FedEx shop. I chatted with a customer service rep about safely shipping the Pyramid. I carefully brought out the piece and asked if the artwork could be safely packed and shipped to arrive by Thanksgiving. He reassured me that he and his partner were experts in packing such quality pieces of art. Before allowing him to pack the piece, I asked him to read the pamphlet I had written that was to accompany the Pyramid on its journey. I wanted him to know how important this special package was and the significance of it. He scanned my words then looked up and said, "Yes ma'am, we'll pack this Pyramid very carefully."

I watched his every move as he gently wrapped the sacred artwork in three cocoons of bubble wrap and used three boxes strategically placed for support and protection of the diamond on the top. I saw the deeper significance of his using 3 layers of wrap, and 3 boxes. He was using great skill and care in the wrapping. I thanked him for not charging us for 3 separate boxes. I knew it was no coincidence that I was directed to go to that location for the shipping of the Pyramid. Soon, it was ready to join the thousands of other packages crisscrossing America this Holiday Season. It was Friday, November 23, 2008.

It was a date I would always remember and honor. I shipped the precious Pyramid, a symbol of honor for all African-Egyptians around the globe. I sent it with love and gratitude, to its intended home, The White House.

That Monday, I received the certified mail receipt confirming that the package was delivered to its final destination. As I read the message, I started to get a disturbing thought. I had mailed the letter of intent to the White House almost two weeks ago. Maybe the Pyramid was delivered before the explanatory letter. It only took 24 hours to confirm my assumption. I received another letter saying that the package did arrive before the letter of intent. I had to act fast.

I hopped in the car and drove to Staples. I needed to fax an emergency notice to the White House informing them of the situation. We were faxing the White House. I think it made the Staples clerk a little nervous. On her first try, she dialed the wrong number. When she re-dialed the line was busy. Turning to me she said, "Oh, everyone must be faxing the White House at the same time as us!" I half-smiled, but was not really amused. I was really worried about the whole project somehow blowing up completely. She dialed a third time and thankfully, the fax finally went through. I just shook my head, knowing that when it comes to this Pyramid, everything has been in threes.

In February of the following year, almost four months after shipping The Pyramid of Love and Gratitude, a mailman brought me a letter from 1600 Pennsylvania Avenue, Washington, D.C. I was so nervous I could hardly open the envelope. The letter stated that the precious gift had been received at the White House. A card of thanks was on its way to me also.

Since those very special days of my life, I have consistently given all the credit to Joseph for the Pyramid being sent to the White House. The entity that gave me instruction for its creation, no doubt influenced Joseph as well and the stained glass shopkeeper. Together we created, duplicated, and carefully delivered a powerful symbol to the seat of power of our country. It is the place where the control of our environment and the source of how our reality is programmed and originates.

Just a few days later, Joseph and I were out enjoying the Florida sunshine, cycling through our neighborhood. When we stopped for a water break, he said, "Do you realize that we sent the third Pyramid?" I really hadn't thought about it until he mentioned it. Of course he was so right. The fact is that the number 3 has come into play on so many different occasions since this all started, I believe is significant. While I sipped my water, my eyes drifted to the beautiful blue sky. I began to envision the Divine Trinity of the Father, the Son, and the Holy Spirit. Then in my minds-eye I was seeing the Pyramid. I felt good. I felt blessed.

The number 3 honors African-Egyptians of the Past(1), Present(2), and the Future(3). It is composed of 3 colors. It was the 3rd Pyramid, wrapped for its journey to the White House in 3 layers of protection, inside 3 boxes.

I have spent many hours researching why this all occurred, to explain why all of this has unfolded, and what it means for us all not just the United States, but the entire world. I now know what this all means, as I write these words in the year of 2019. The pyramid serves all humanity by bringing great salvation to all people. It gently assists in removing the layers of burden from life cycles passed, spent in false realities of doom and chaos, keeping us from ever rising above. It represents and empowers Love and Forgiveness. As the veil of the traditional system is now lifted, we reflect on what has been taught and practiced, in order to hold all of society in a false reality. Our time is finally now to begin exercising critical thinking, by acknowledging the actions of those in whom we trust to lead.

NOTES

73

7

THE MEMORY OF THE SUBCONSCIOUS MIND

Without understanding our collective past, we are subject to attempts to hold all of society in a false existence of reality. That can create financial situations that punish the world's underclass and, to a lesser extent, those who enjoy a reality and believe they are part of a privileged group. I'm certain that one of the finest qualities of my relationship with Joseph was our regular, deep conversations. We thoroughly discussed how real life situations can cause a "domino effect" eventually affecting all of humanity. During one particular chat, while taking a water break during one of our regular bicycle rides, he explained his interpretation of the human mind. I was fascinated. He described his theory that our brain was like a huge vault, constantly storing more and more data. He thought that our nighttime dreams are often triggered by past experiences. A childhood event, good, bad, or whatever, from our past life, could possibly trigger our dreams.

The conversation really fired up my memory! That evening I vividly recalled an experience when I was 9 years old. I realized that he was absolutely right! The past images streamed back. It was 1974. Federal courts were beginning to enforce desegregation across the country. That summer, new school districts were being rezoned. Some African kids from across town were reassigned to the school I had attended since kindergarten. I was in the fourth grade at Kaley Elementary. Forced bussing would begin just a few months.

I recalled that as the time approached, I felt an uneasiness building inside me. I didn't know why I was feeling that way. Off to school I went on day one. For the first time in my young life I was to sharing my school with children living miles away, arriving on buses, and each one of them a different color than me and my friends. I had never really been around "these particular children" before, although I heard plenty mean-spirited talk about them from grownups, including my parents. I didn't know what to expect.

My parents spoke of "colored" people. I assumed that meant the colors of my crayons; red, yellow and blue. When the new kids arrived and got off the buses, I could plainly see that they were darker-skinned than I was, but not red, or green, or even purple! I was almost a little dis-appointed that they weren't.

Within a few days I made friends with a small African girl named Carolyn. She was so smart and very pretty. Soon we were best friends. Unfortunately, not all the new classmates were as nice. Several months later, two African boys started picking on me. They mocked me and called me hurtful names, even though I had not done anything to them. I just couldn't understand why they didn't like me. Again, I hadn't done anything to them.

The verbal abuse continued for several weeks and months, but I was afraid to tell my parents, especially my Dad. My father was a very complex man who believed that fighting was a good way to settle disputes. In his mind, winning was the only option. As a youngster, I remember him lecturing my older siblings that if they were in a fight at school and lost, then he would kick their behinds when he got home from work. I had no intention of fighting the boys although I knew how to fight. I promised myself to only fight for self-protection.

The name-callers kept up their verbal barrage every day, slowly breaking me down inside even further. It was really getting to me, mentally. My fear was getting stronger and stronger. I couldn't take it anymore! I decided to confide in my mother, but only after she swore not to tell dad. I told her the whole story. Two African boys at school were calling me hurtful names even though I had not done anything to them. Mom listened intently then said in a calm voice, "Oh, they are just mad about slavery."

"What is slavery?" I asked.

Slowly, mom began to explain what that awful word meant. She told me that the boys' ancestors were probably slaves, here in America. It felt like a knife stabbed my heart because I knew she was telling me the truth. I felt ashamed to be born to such a race that was involved in this. As a child I had no other context, yet my heart was breaking as I shared a pain I had never felt before.

I began to recall my father's physical and emotional abuse of me and my sisters and brothers. I thought that it must have been like that for slaves all those years ago. From my childish perspective, it was the same thing. I lived in an abusive home. I was treated no better than a servant. As I grew older, and my siblings were no longer at home, I was regularly targeted if I failed to do what I was told. Often, I was harassed for no reason. Because of the constant threat of abuse, I became a child that acted out of fear. That fear became my reality. My elementary school was once my safe haven, but not anymore. Unknowingly, my own reality of fear and anguish of my home life led me to draw even more trauma to my childhood.

The boys that were hateful towards me at school had been taught to hate white people, out of fear. They had experienced threats of abuse and violence their entire, young lives. They also were acting out of broken hearts, filled with fear and repressed anger. My heart wept for them. They were simply reacting to me in a manner that relieved some inner, constant pain. Although I pitied their ancestral history, and the scars it left on their lives and all past generations, I was determined to not be the target for their rage any longer. Soon, somehow, it had to end.

The morning after my mother's "slavery" talk, I was up early and off to school. I really hated seeing those boys again. I thought about skipping school, but soon thought better I knew my father would beat me badly if I got caught, so that was not an option.

I began a silent chat with God. I spoke out loud that I wished I could do something nice for black people, to somehow make it better. I knew that words of apology alone would not suffice. I told God that I could handle their name-calling and that my shoulders were broad enough, because anything was better than a "whipping" from my father.

With peace and love in my heart, I released my fear and went along to school, not thinking anything would come of it. Over the next several days, the boys continued to call me names but now I understood why they were acting out. I kept trying very hard not take it personally. I did my best ignoring their taunts. I was handling it all, until one day, when the words became physical violence.

One day, while walking down the hall, right after lunch, and without any warning, one of the boys crept up from behind. He pushed me in the back so strongly that my neck whiplashed violently backwards, it really hurt! Something seemed to snap in my mind. I instinctively went into self-protection mode. I went into a total rage. I reeled around with swinging fists, landing a solid punch to the boy's face. It shocked him as they both backed away from me quickly. I could see their fear and shock at my reaction. I saw the beginnings of a pretty good bruised and blackened eye developing.

Within seconds they turned and hustled away into the hallway crowd. I thought I saw them head for the Office. I was standing there alone, feeling sad for what had just happened. When the office door swung open, I knew who was coming to see me. I had to explain to the school principal, Mr. Coffman, what had happened and why, from my perspective. I told him as much of the story as I could remember, not embellishing anything, but not backing away from the truth.

What frightened me most, was the thought that my father would be called and he would find out that these boys had been verbally abusing me for most of the school year. I know I would get beaten severely from my dad. Fortunately, that call was never made. Instead Mr. Coffman brought the boys and me together in his office. We all sat down and talked. The boys eventually apologized and I said I was sorry for delivering the shiner. As best it could, it worked out. In the days ahead, that well-thrown punch appeared to do the trick too. The boys never harassed me again. The year would end peacefully. The school doors opened to another Florida summer.

I remembered how I had said a prayer, asking God to help me, to show me a path to stop the boy's mistreatment. They did stop bothering me. My father never found out. I was spared a painful beating. Normally I didn't really expect God to answer my occasional prayers. Did he or was it just coincidence? The spiritual world was a confusing mystery to my young mind, to be sure.

During those days of my youth until I reached a milestone at age 42, I felt that there was an important part missing in me. I lived with a strange feeling that something was constantly in the shadows of my existence, trying to become a part of me in positive ways. I hadn't learned to open the doors or windows of my soul. Often then I also experienced dark thoughts. My mind would fill with notions that my existence was a mistake, I would brood over why I was born to live a life without love and support. Over and over I would wonder why my family existed as it did, with dysfunction and violence, and how that was stealing any youthful joy I could imagine. It was only much later, as an adult with a family of my own, that I realized that my life, regardless of its trauma, was not as "unique" as I thought as a teen.

It was not until my eyes and heart were opened, years later, that I understood that I was chosen to go through those awful years to be a living example to others of the power that is within us all. I learned by living that we control, within each of us, the most, incredible piece of equipment the universe ever created, our Mind. It begins simply with belief in oneself. Give it a try. You'll see what I mean. Your best life is waiting for you. Live, love and be happy. Being full of bliss is a universal intention for us all.

Through my spiritual confusion, I often wondered which god we were supposed to be following. I now understand that stories are often dreamed up to keep the populace in a certain reality of beliefs and practices. Any group, religion, or government seeking to maintain control and authority will create their narratives to achieve those goals. I discovered from a close self-examination of my own life's journey. I would make progress, but then find myself in a never-ending cycle of behavior. I never knew that I was capable of achievement through thought and reason, and by relying on Science. I witnessed many others in my same situation, unable to move their spiritual world to a higher truth, shackled to someone else's rules or ideas. I thought, why have a mind, if it is constantly limited by others? Do we exist in such limitations? Why are some people capable of success, while others remain merely servants?

I also thought a great deal about African-Egyptians. What has caused them, throughout history, to be victimized so profoundly? Through my self-discovery and research I know that it has never been a skin color issue, or their physical appearance. It wasn't ever visible to the naked eye. The reason for the fear and hatred leveled at those proud cultures is what is on the inside of their DNA. They hold it all! The imposters, who have ensured that African-Egyptians are repressed, remain in control of programming all cultures by creating and teaching false dogmas, to ensure compliance and loyalty.

A major milestone in my life was finally understanding those ancient connections. It is the most important lesson any of us can be taught. It strips bare the falsehoods, lies, and deceptions we are all coerced to believe. I discovered that I was missing the number one element that creates the solid foundation of understanding we all seek. It is Love! Its impact on my spirit while in this physical form defies description.

NOTES

86

87

FINDING MYSELF THROUGH THE PYRAMID

At this book's re-publishing, it's been almost twenty years since my important mission began. I believe that my life is to serve the Higher Realm. As a nine-year-old, desperate, little girl, I vividly recall looking skyward on a sunny Florida morning and sharing with God, I wish I could do something nice for the African people. I believe that entreaty led, all those years later, to that amazing dream and my calling, to create *The Pyramid of Love and Gratitude*. Years later, I discovered my innate ability to astral travel regularly. Now, I can reconnect to my powerful past and receive specific messages and directions from the knowledge stored in my cells.

I know my past is profoundly connected to the legacy of African-Egyptians. I understand their plight of enslavement wasn't connected to their skin, but their DNA. In order to successfully enslave the human reality, those who stood to benefit attempted to erase the knowledge in their cells. If they could achieve that, a reality to control all peaceful beings could be attained. This brought about the mass enslavement that darkens all our histories. Africans were the first chosen race. They once taught great alchemy and the importance of love and gratitude. Their Knowledge is eternal and must be preserved, protected, and understood by all.

I have preserved many of the messages I received in my original astral travels over a decade ago. I would like to share a few…

The voice spoke, *"You are to create the Pyramid of Love and Gratitude in honor of Africans of the past, present to eternity."* (Initial life-changing message).

"Only when you open your heart to the love that has been waiting for you in the universe; only then will you understand the true meaning of life and living."

My angels returned to me to say,

"Unity is happening right now, do not let it worry you."

"Cut the gray strings that bind you to the past.

Your mind holds all the data to all of your past lives."

"Study all the words that start with the letters "re'.

These words are in harmony with the universe."

"You must trust in who you are and love yourself enough."

In order to fix what has been wrong in your whole life."

"The secret to freedom from everything is within your state of mind.

Your reality is the key to freedom."

I realize now, more than ever, that my state of mind and my thoughts are the driving forces for all that happens to me. The seeds of experiences, planted in my mind as a child, taught me what to think and believe. I know that their harvest held me back, even to adulthood. Only by becoming enlightened and aware of my true self as a powerful being, was I able to progress in life.

Our pathway is to dive deeply into what is real. That is your inner self. Ignore thoughts that your shortcomings are "a genetic thing" or "that you were born into the wrong gene pool". They are simply untrue. Believe in yourself and leave your cocoon, becoming a forgiving and loving human being. Just open your mind. Eliminate self-doubt. Be strong! Stand your ground! Don't accept failure! You have the power to reconstruct your mind into success. You are worth that and so much more. We are a network of learning and growing subjects.

93

94

9

WHY THE PURPOSE OF THE PYRAMID RESONATES IN MY HEART

As I've made plain in the preceding chapters, although I love and respect all peoples of the Earth, the African people, having originated from Egyptian beginnings, hold a very special place in my heart. Even though I am Caucasian, I have always felt a profound connection to the innermost part of their energy. Their strength and their cultural pride, in spite of centuries of forced enslavement, is unmatched as one studies ethnic/racial struggles throughout history. Their systemic repression that stretched continents never defeated their inner strength. They have maintained abiding faith in themselves and in the higher power. The time has come for African Egyptians, worldwide, to telepathically connect their cognitive minds to one another, and bring Love and Gratitude back to the forefront of our reality.

As a child, I had an uncanny feeling that their history was vast and long. Setting them apart from other cultures was their spirit and soul. When I became aware and enlightened, I fully understood their indomitable inner strength and sheer determination in the face of any struggle. From their earliest beginnings along the savannahs of the great rivers Congo, Niger, or Zambezi, African culture has been in perpetual communication with the creator of the universe.

The terrible period of slavery in the European colonies, and eventually to this continent, began with the documented arrival of the first slave in North America in 1654. From those earliest days onward, Africans slaves and their families were denied any formal education while enslaved. The plantation owners foolishly thought that by keeping their "property" illiterate, their control over the Africans could never be challenged.

As the slaves would gather at nightfall, after another grueling day in the cotton fields, they would meditate with movements of dance and by chanting. Speaking in "tongues" was their private way of communicating with their higher power and with each other. If their lives were physically shackled, their minds were not. The Africans of the Americas gained complete freedom of their souls.

American President Abraham Lincoln will be forever known as the Great Emancipator, by extending love to every African alive. The famous Proclamation, passed by a contentious Congress, represents as the most important legislative action ever taken in America. It granted the slave unfettered freedom. It was viewed by Southern, slave-owning people, as treasonous and enough to cause secession from the Union by eleven states.

As the Civil War raged on for four bloody years, Lincoln wrote that he saw the conflict "as a morale crusade to wipe out the sin of slavery." In a letter to H.L. Pierce dated April 6, 1859, two years before the first shot was fired, Lincoln made his feeling known writing, "He, who would be no slave, must consent to have no slave. Those who deny freedom to others deserve it not for themselves and, under a just God, cannot long retain it." Murdered in 1865 by an assassin's bullet, he gave his life to repair the damage brought on by 200 years of slavery. The long national nightmare finally ended in 1865.

I have long believed that the African people are the Chosen ones to guide humanity toward enlightenment. To be Chosen means never forsaking the laws of the Universe which created us. To be Chosen means knowledge from the beginning of time had been recorded in their cells and can be shared with all humanity. To be Chosen means to not put material items ahead of Love for all of humanity. To be a Chosen means to not create trauma in the lives of others, clouding what proper reality should be. Chosen people are not braggarts. They humbly serve humanity with love and kindness. A Chosen one understands that nothing is ever really owned. The land doesn't belong to anyone. I pray that the knowledge conveyed in this message has touched a profound place in all the hearts of humanity. Exist in a reality of love and truth. Be mindful of your experiences in order to reach the correct knowledge that advances us to exist in countless higher dimensions of reality.

10

A BEACON SHINES FOR ALL THE WORLD

The creation of the *Pyramid of Love and Gratitude*, and its purpose in the world today, have had a profound impact on my life and all others who are chosen for this grand purpose, allowing their minds to open their hearts. That act of self-discovery connects our hearts to the place where all love and splendor originate.

As the years continue to accumulate since the journey of this true story, I have continued to evaluate my life's experiences. I fully understand the importance of my being guided to create not just one, but three Pyramids. I have gained more insight over time to its divine purpose for us all and it's being an entryway for the higher realm to penetrate this reality without breaking Universal Laws.

I am a willing vessel, assisting in this realm of existence to help the higher realm and all of humanity to reach success. The Pyramids of Love and Gratitude are all here to serve as beacons, guiding all who will allow its energy to flow within them to know deeper and more fulfilling love, wisdom and kindness.

I would have never imagined, prior to this life, the importance for us all to use our minds correctly, guided by our hearts, to create the correct reality. We must seize this opportunity of life! We must start thinking and acting in kindness to others, regardless of what inappropriate acts occurred against others. We must improve our reality now. We must save our energy of consciousness while in a living consciousness state. We are required to harbor great comprehension in our cells of experience of Love to all and the greatest respect to all people.

My miraculous discoveries have led me to know with certainty that The African is descendant of the Egyptian civilization, and are the true chosen ones to lead this consciousness in a reality of Love and Gratitude. I fully understand that Love is our greatest power and without it, our reality is doomed. Our ability to be aware of what is happening is our greatest success. Our reality is improving with the correct elements, the number one is love.

I have expressed my innermost experiences and thoughts. We all must be keen in our ability to think and understand. We must remain several steps ahead of the agenda of those who would hold us to the false reality by wasting our precious time.

There have been some encouraging signs all around the world. I was so excited with the election of our first African President, Barack Obama. Africans taking all sorts of positions that were previously denied to them. I really felt so good about this knew equality. Things seemed to be going so well, but the good feeling was short-lived. I now see it all as lies, distortions and self-serving of the controlling reality. Now the puppeteers are using the Africans Egyptians to assist them in continuing their purpose. Although systematically held down for centuries, Africans are now being placed in many government positions. A system that sacrificed so many to slavery's cruelty, now uplifts. I only wish it weren't true. During the post-Civil War Reconstruction Period, Africans were placed in offices in many levels of government, while being controlled by the carpet-baggers. This is a period of similar deception.

In 2019, a young African television actor faked a racist, violent attack. Prosecutors found evidence that he, in fact, had created the incident to pander to his mostly African fan base. Instead of prosecuting the case, he was cleared of all charges by the Prosecuting Attorney for the county, an African lady. It was painfully obvious the decision was made as they were to cooperate with a higher council over them. The controls are still firm. The real ruling parties still firmly have the reins. Of course, Africans must be in positions of civil authority, just not under anyone's secret control.

With the events of the world unfolding at lightning speed, it's easy for us all to feel some degree of insignificance, especially if we presume that we can change the course of things. If you let Science be our guide, you can impact the most important soul you know…you. Try this. Each person's life is for a specific amount of time. Make a list. Jot down all the things that impact your reality. List them all. Be open with your observation of what you fear and why? When you've finished, evaluate the list carefully. Do you want to cross over with the taught reality of what we fear and hated? Do you know that most people cross over with the burden of their captors? It is true. You can re-write that list. You can shed the old thoughts and barriers. You can free your spirit. There is no such of a thing as death. We transcend to a place of pure reality.

NOTES

11

LOVE IS THE ANSWER

The binding element is the Science of using both mind and heart together. Working as one, they form and create all our desires. To believe in oneself forced me to love myself. If I ever doubted who I truly was, that way of thinking existed only due to what I had been previously taught. I realized my thoughts would become my experiences. It was mind-boggling to grasp how all the pieces fit together so perfectly!

My research led me to ask many questions to people educated in the natural laws of physics. Unfortunately, I never seemed to get satisfactory answers, chiefly because those I asked did not know their inner selves. They couldn't apply the Science beyond the textbook's page. I realized that my baseline question was, "how do things become to be "created? What makes all matter form into a particular object and shape?"

One night, out with friends, the answers just flew into my mind. I realized that atoms and their extended elements, were floating effortlessly around us all constantly, invisible to the naked eye. How they form to become objects out of the elements provided by the universe requires a command from the Cognitive Mind. Without that, atoms just float randomly, in chaos. The mind must know exactly, without any doubt, what will appear. This was a major breakthrough for my own understanding! I now had the formula and necessary elements to create whatever I wished.

As I thought more and more about these revelations I came to see a simple truth. The foundation to all creation requires the element of love. My life was changing for the better. I witnessed all that was happening in the world with fresh eyes. I understood that we are all programmed, like a computer application, to exist in a lesser reality than we were capable of achieving. I could easily see that wealth, status or entitlement had nothing to do with our life's true purpose. Without love, our life experience is meaningless.

We are not alive on this planet to destroy each other, or to war against other nations or to judge each other based on race, color, creed, or physical limitations. Sadly, I believe that our programming does not uphold those tenets, thereby allowing each of us to reach for the higher levels of knowledge. So many live in a state of constant threat and fear, as simple survival is the singular, daily struggle.

If we could only practice peace and love throughout the lands of the Earth. What a beautiful world this would be! I believe it is what we must all strive for, in our daily lives. I do, and there are millions of individuals just like me, existing in a reality of love and peace. Love is inside us all, buried deeply in many, in the catacombs of our past lives. Acts of kindness awaken that precious element. We bring it into this existence. Deliberate actions of others, and the agendas they promote, can smother Love from being expressed or experienced. Hurtful experiences mold our reality. The attraction of material wealth or assets, can rob people of something far more fulfilling, the ability to love or be loved unconditionally by the energy of all others, created by the human heart.

I have had the opportunity to experiment with what I have discovered. In spite of what Africans have endured by all governments around the world, the love is still alive within their spirits, flowing out, creating a proud reality. I pray that everyone will see their own invisible atoms and envision a reality for everyone on this planet to experience love and peace. Our minds together can do this, connecting all humanity to a higher realm of bliss.

Each of us can begin the small steps toward a better reality for all. Laying the bricks for the foundation of Love starts with kindness to others, regardless of circumstance. Always remember,

Other people may have experienced horrific events far worse than anything we might imagine. Horrific experiences from past lives might be repeating in their present lives, dragging down the spirit through eternity. Others may be bound to a religion that limits individual thought, forcing a way of life on them, damning a greater purpose.

Those existing in a mind of peace and love must not become angered by others who were forced their entire lives to exist in such a state of mind. They are victims not enemies. The love and kindness is given to all unconditionally.

Showing others love can break their curse of darkness. Everyone deserves love and true freedom. We cannot allow oceans or walls to divide us. The Universe is vast. Our contract with it cannot be broken. We are here to learn correctly and we are here to grow and achieve. If I know a better way to help myself and others, I will share it with my fellow man, woman, or child. Together we can change our reality to something far grander than our wildest dreams.

Will true and eternal peace happen in my life cycle? I know it will! I had to understand what needs to change in all our lives in order make this a part of my journey and bring us all to a better world. The old ways, from the beginnings of Man, must cease. The simple way we are taught as children, to listen and believe our elders, to accept the dogmas of hate and war and conflict as more than normal, to memorialize and praise the victorious outcomes over vanquished enemies…for what purpose? Are our lives improved and made better? Are we safer? What do the pure, little minds retain from these stories and lessons except fear and darkness? The latest conflicts only act to program our future generations. This only repeats and repeats becoming a horrible, sick cycle. It seemed as if my entire 8th grade history class was about the Civil War. Understanding slavery's history in America was important, but not the facts about four years of battlefield slaughter from Bull Run to Gettysburg. Today students are taught about 9/11, and asked to share how that tragedy makes them feel? When they see video of people leaping to their death from 100 stories above Manhattan, how should they feel? The exercise is ludicrous. The point of the teaching, better called indoctrination, is to keep the "old ways" pumping with fresh blood.

When that child leaves the schoolhouse and begins to head home, she might be pedaling her shiny red bicycle. That simple action, of moving those pedals in that circular motion, over and over again was probably taught through "trial and error". A few scratches and bruises helped enable the great human mind to do its best work, to remember and repeat. We are all capable of doing this basic function. We are not limited to self-teaching how to use a mechanical device. No, we are able to do so much more. When we have thoughts and can project the outcome of an action we actually have the power, within our bodies, to make it reality! We build cells that become lasting parts of our life-structure. We not only build cells, we cause them to expand and spread, molding a new reality for the new experience. If you understand the functions of the mind, and its powerful abilities, it is only a matter of time before your thoughts, good or bad, become a true experience in the reality of our own creation.

We can create our own reality. Many use their mind to envision and create wealth, successfully. However, without connecting the wealth to improving humankind, the threat of corruption and darkness will linger and burden our realities. These burdens mold a reality, recorded in the cells. This absolutely isn't what we are here to experience. We are here to have a physical experience, while expanding our ability to love each other and create by the abilities of the mind utilizing the elements provided free of charge by the Universe.

We are a network of energy inside physical bodies. We have learned to communicate with other like-beings. Together we achieve, live, and learn. When we think, those thoughts record in the cells of our DNA, invariably affecting our reality. We all are a collective network of energy that can work together in harmony to get our experiences completed.

How our thoughts have a direct impact on our lives can be traced through our history. I invite you to reflect on the years of your life that are past. Do your best to recall ideas or dreams, even visions of the future. Think about how any of them have come to fruition. Those events or desires of your past that have become reality were only possible because you thought the concepts and allowed them to be recorded in your cells.

In my journey, I have learned to not see people as a physical body, but to see a spirit of energy. This allows me to have no bias with race, skin color, creed, education, etc. Those discriminators only act as reminders to keep our energies divided. I believe that Love keeps us in a reality of being open and receptive. I know that by each wholesome thought we send out, it is working to change future realities. As we engage our minds to think properly, while opening our hearts, we embark upon a new era of existence. Every human we encounter is due unconditional love.

In my previous book, I explored my self-discoveries, including my views on religions and belief systems. This book has chronicled my personal mission to create The Pyramid of Love and Gratitude, as I was compelled to do by my spiritual guidance. My unique experience has brought many revelatory images to my mind's eye. I am convinced that we cannot continue to exist divided by religion, cultural exclusivity, or where we live on this planet. We have to live together on this blue marble in the vastness of space. The Earth expands constantly, in reaction to an ever-expanding Universe. It is our home to share and cherish. Look at our beautiful planet! It is so beautiful! Consider the diversity of the land, its natural life, the animals, the oceans, and all the beautiful people. There is so much love around, and in us all.

The opportunity is now for us to change forever how our lives are captured by an enslaving reality. The efforts we make today, will benefit us all along our journeys tomorrow. Prior to our enlightenment, the world's elite and their offspring had created the false concepts of control. When we are forced to exist under such conditions, we all are cheated out of own harvest and fruits. When we pass, we also leave this consciousness with their practices and realities still filling our energies, infecting our essence for the next life cycle.

Now is the time to dig in and make changes that will absolutely be rewarding and will change us forever! If it feels bad don't do it! If it enlightens you, then you are on the right track. These are life lessons. Know in your heart that we are not chained to anyone's reality. Serve the greatest purpose, which is Love. Today a wave of love, knowledge and peace, blankets our reality for the betterment of all. Love is our highest purpose. It is the firmament to all our endeavors. We are born to this earth-home to choose our experiences and strive for them, while growing in a loving, spiritual way. The Pyramids I was directed to create, are the portals of the higher realm. They help to connect us all to the past, present, and future. They are true beacons, expressions of witness to those whose existence in this world serves to provide a shining example of the power and majesty of Love we are all capable of manifesting, if we only will see, hear, and feel with our hearts.

123

ACKNOWLEDGMENTS

This book would not have been possible without all the individuals touched by this wonderful experience in the making of the pyramids. In order, I would like to share with you the chosen ones who were placed in the path of the pyramid. Risa Reynolds, my best friend and the first editor of this book, without her I would have not gone as far as I have. Risa believed in my experience and was a part of my life while it all unfolded and she too shares the passion of the message behind this beautiful piece of art and the journey behind the symbol of love and gratitude. Acknowledgement to Terra Jarvis for performing the second round of proof reading.

Finally the greatest communicator, Chuck Gillespie, who has performed the third and final editing of this book. My love to Tammy Pearce for being a spiritual guide. I also wish to thank my family starting with my daughters Crista, Michelle, and Tiffany. Following, sons Darrel Clay and Charles Norris then the granddaughters Samantha Asha, Arianna, Cheyenne and Kharlee, grandson's Robert, Brandon and R.J, my mother Joann Grauberger. Next, my sisters Elizabeth Thompson, Joanna Jensen, Melody Pearce and Anabel Johns and my brother Wade W. Pearce Jr., and cousin "Lil" 3 Brittany. I also thank my extended family Joseph Rivers, Maxx Cyber Rivers, Yolanda Rivers, Joe Rivers. Diane Davis, thank you for being one of Cassadaga's most gifted mediums. I could not have had my dreams realized without the talented efforts of artist Rich at Atlantic Stained Glass; Sarah and Mark at Stained Glass Gallery, Duane at Logo's, Mike at FED-X, Andrew at Tags are Us, Darcey at Boomerang Graphics. Also, my close friends, victorious Vikkie Hankins, Cilla, Frank, Dwayne, Samuel, Anna and Camille – thank you for your encouragement.

Finally, I wish to recognize, in loving memory of my two dads: my biological father Wade W. Pearce Sr., and my "second" father Harry V. Grauberger, my mother Joann Grauberger, as well as my brother Clarence W. Pearce.

I want to express my complete love for both my parents and I do not hold them in any way responsible for how they raised my siblings and me. For if they had been taught differently, I would not have been able to share with you what I now know to be true. So, thank you mom and dad, I love you both very much!

I open my heart of gratitude to you all for believing in this divine message. Without you all, none of this could have ever been. You all have opened your heart to love that has been waiting for you in this wonderful universe in which we live. Thank you from the bottom of my heart my brothers and sisters and our holy spirit in God's name, AMEN!

AFTERWORD

CRISTA ANN MCLEOD

An entourage of thoughts and desires, combined in a series of revelations that conspires to bring us closer to the dawn of a new nation, to expand the horizon and show the destination.

This is the time we must all reflect on our tribulations and proceed through with no provocation.

Leading the world through oppression and depredation and healing the effects of mass creation All of these things bring the hope for mankind, to pursue our true purpose in life with a passion. Forgiving our past through procreation, rehabilitating humanity with appreciation.

Man, woman and child alike, all races, all cultures will supersede this plight. We have all helped to create this magnificent light of inspiration. All roads lead somewhere and it is there where peace and happiness dwells, within you.

SOURCES

1. Freedman, Russell. Lincoln A Photobiography. Carion Books, 1987, p. 5.

2. Ibid, p. 135.